NEURALINK IN TESLA OPTIMUS BOT?
Everything You Need to Know

The Power of Thought Meets the Precision of Robotics—The Future Is Here

J. Andy Peters

Copyright ©J. Andy Peters, *2024*.

All rights reserved. No part of this publication may be reproduced, distributed, or transmitted in any form or by any means, including photocopying, recording, or other electronic or mechanical methods, without the prior written permission of the publisher, except in the case of brief quotations embodied in critical reviews and certain other noncommercial uses permitted by copyright law.

Table of Contents

Introduction 3
Chapter 1: What is Neuralink? 6
Chapter 2: Tesla Optimus: The Future of Humanoid Robots 19
Chapter 3: How Neuralink and Optimus Work Together 30
Chapter 4: Robotics and Human Mobility: Restoring and Enhancing Abilities 48
Chapter 5: Neuralink and Tesla: Shaping the Future of Human Enhancement 61
Chapter 6: Ethical, Social, and Security Implications. 73
Chapter 7: The Road Ahead: Neuralink, Tesla, and the Future of Humanity 92
Conclusion 111

Introduction

In the not-so-distant future, the boundaries between human capabilities and machines will become increasingly blurred, with groundbreaking technologies pushing the limits of what we thought was possible. At the forefront of this revolution is Neuralink, a neurotechnology company founded by Elon Musk, and Tesla's humanoid robot, Optimus—two innovations that, when combined, hold the potential to reshape the way we live, work, and interact with the world.

Neuralink's mission has always been to create a seamless connection between the human brain and technology. With its brain-computer interface, or BCI, the company has made strides toward turning thoughts into actions, offering hope to those with neurological impairments and disabilities. On the other hand, Tesla's Optimus humanoid robot represents the next evolution of artificial intelligence and robotics, designed to replicate human movement with incredible precision.

Together, Neuralink and Optimus are more than just two separate technological feats—they are a fusion of human mind and machine, pushing the boundaries of what can be achieved.

This collaboration isn't just about restoring lost abilities or creating the next wave of intelligent robots. It's about enhancing what it means to be human. Imagine being able to control an entire robotic body with nothing but your thoughts, or regaining mobility with a prosthetic limb that responds exactly like a biological one. These possibilities aren't just hypothetical anymore. They're happening right now.

The aim of this book is to explore how this fusion of brain-computer interfaces and robotics is not only going to change the way we interact with technology but how it will redefine human potential. From restoring lost mobility to enhancing physical capabilities, the applications are vast, and the implications are far-reaching. This journey will delve into themes of human enhancement, the rise of advanced robotics, and the role these technologies will play in shaping the future. What was once science fiction is fast becoming reality—and as we stand at the edge of this new frontier, the possibilities seem endless.

Chapter 1: What is Neuralink?

Neuralink's journey began with a bold vision—to merge the human brain with computers, essentially creating a seamless interface between our biological mind and artificial intelligence. Founded in 2016 by Elon Musk, Neuralink aimed to tackle some of humanity's most significant neurological challenges by enabling direct communication between the brain and machines. Musk's goal was not just to develop cutting-edge technology but to address the fundamental limitations of the human brain. He envisioned a future where brain-computer interfaces (BCIs) could help treat neurological diseases, restore lost functions, and, ultimately, enhance cognitive abilities.

At the core of Neuralink's vision is the belief that the human brain is not just a biological organ, but a supercomputer capable of so much more. But despite its vast processing power, the brain remains largely isolated from the digital world. Neuralink's mission is to bridge this gap and create an interface that would allow our brains to interact directly with technology. The company's long-term goal goes beyond medical applications—it envisions a future where humans can use BCIs to enhance their cognitive and physical abilities, keep pace with advancements in AI, and even overcome the limitations of our biology.

The N1 implant, Neuralink's flagship brain-computer interface, is a game-changing technology designed to make this vision a reality. It's a tiny, wireless device that is surgically implanted into the brain, targeting the area responsible for movement control. The implant is equipped with thousands of tiny electrodes that can read neural activity, pick up on electrical signals, and translate them into actionable commands that can control machines. It's essentially a bridge between the brain's electrical impulses and external devices, translating thoughts into actions.

The N1 implant's primary goal is to help people with neurological impairments regain control over their environment. Initially, Neuralink focused on helping individuals with conditions like paralysis regain the ability to interact with digital devices. In its first successful demonstration, participants were able to control computers, smartphones, and even play video games purely by thinking. The success of these early trials demonstrated that the technology could decode brain signals with remarkable accuracy, opening up a world of possibilities for those with disabilities.

But the potential applications of the N1 implant don't stop there. As Neuralink continues to refine the technology, its capabilities expand beyond just digital interaction. The ultimate aim is to allow users to control robotic limbs, even entire robotic bodies, with their thoughts, offering a new level of mobility and independence to those with disabilities. In the coming years, this technology could also be adapted for broader uses, such as improving cognitive functions or enhancing human abilities to interact with increasingly complex technologies.

What makes Neuralink's N1 implant revolutionary is not just its precision, but its potential to change the way we think about human capabilities. By connecting the brain to machines, Neuralink is opening up new possibilities for human enhancement, offering a future where we can seamlessly merge biology and technology in ways that were once thought impossible.

Neuralink's N1 implant is a marvel of modern neuroscience and engineering, designed to connect the human brain directly to the digital world. At its core, it functions as a brain-computer interface (BCI), enabling the brain's natural electrical activity to control external devices. The technology taps into the brain's intricate network of neurons, which communicate through electrical signals, to translate thoughts into commands that can be interpreted by machines. The process starts with the N1 implant, a small device that is surgically embedded into the brain's cortex, specifically the motor regions responsible for controlling movement.

The implantation procedure is precise and minimally invasive. Rather than relying on traditional surgical techniques, Neuralink uses a specialized robotic system capable of inserting thousands of tiny, flexible electrodes into the brain. These electrodes are thinner than human hair, allowing them to penetrate the brain tissue without causing significant damage. Once the electrodes are in place, they are able to detect electrical signals generated by neurons. These signals represent the brain's intentions, whether it's thinking about moving a limb or focusing attention on a particular task.

The N1 implant processes these neural signals and converts them into commands that a computer or robotic device can interpret. Essentially, it allows the brain to communicate directly with machines, bypassing the need for physical interaction. The connection is wireless, meaning there are no bulky wires or external attachments—just a sleek, unobtrusive device that seamlessly interacts with the brain. The result is a system that can convert thought into action, making it possible for individuals to control technology with their minds in real time.

This remarkable technology isn't just theoretical—it's already being tested and refined to help those with neurological impairments regain a level of control over their lives that was once unimaginable. For individuals who suffer from conditions like paralysis, the N1 implant offers a way to interact with the digital world in ways that were once out of reach. Using the N1, people who have lost the ability to move their limbs can, for the first time, operate devices like smartphones and computers purely with their thoughts. The system detects the brain's motor signals and translates them into commands that control digital interfaces, allowing users to scroll through a smartphone screen, type messages, or play video games without using their hands.

This capability represents a huge leap forward for those living with disabilities. For example, individuals who've lost the use of their hands due to spinal injuries or neurological conditions like ALS (amyotrophic lateral sclerosis) can use the N1 implant to regain the ability to control basic functions on a computer. In some early trials, participants were even able to play video games hands-free, using only their thoughts to manipulate the game interface. This was not just a theoretical demonstration; it was the first glimpse into a world where people with disabilities could experience a level of independence previously thought unattainable.

The significance of these early uses cannot be overstated. In just a few short years, Neuralink has moved from the realm of speculative technology to the reality of clinical trials and real-world applications. The N1 implant's ability to help individuals with neurological impairments regain control over digital devices is just the beginning. As the technology continues to evolve, its potential applications are limitless, ranging from assisting with mobility via robotic limbs to perhaps one day providing new ways of interacting with the world around us.

In this new age of technology, the N1 implant is a pioneering step in merging human thought with machine precision, opening doors to a future where the brain's capacity to control machines is no longer a distant dream, but an everyday reality. And as this technology continues to advance, it holds the promise of transforming not only the lives of those with disabilities but the very nature of human interaction with technology itself.

The Convoy Study represents a significant milestone in Neuralink's journey and a pivotal step toward realizing the full potential of brain-computer interfaces. As Neuralink continues to push the boundaries of what's possible with their

N1 implant, this groundbreaking study aims to take the technology beyond digital devices and into the physical world. In simple terms, the Convoy Study is designed to show that the brain can control not just a smartphone or computer but advanced robotic limbs and other mechanical systems. It's one of the first trials to demonstrate that thoughts alone can operate complex robotic bodies with the same precision as biological limbs.

The study itself builds on the early successes of Neuralink's previous research, where the N1 implant was able to decode neural signals and use them to control basic digital devices. In the past, these demonstrations showcased participants controlling cursors on screens, sending emails, and even playing video games using only their thoughts. While these were monumental achievements, they were still limited to interacting with digital, non-physical objects. The Convoy Study, however, aims to extend that control to the physical world, focusing on the integration of brain signals with Tesla's humanoid robot, Optimus.

Tesla's Optimus robot is designed to replicate human movement as accurately as possible, using an advanced combination of AI, robotics, and human biomechanics. With its lifelike dexterity and precision, Optimus can perform tasks that range from simple motions like walking and picking up objects to more intricate actions such as manipulating delicate materials. By connecting the N1 implant to Optimus, Neuralink's goal is to show that a human mind can control this robot just as naturally and fluidly as it would their own limbs.

The importance of the Convoy Study goes beyond just advancing technology—it marks a fundamental shift in how we think about human-machine interaction. Until now, we've only been able to control robots or prosthetics with manual input, whether through joysticks, buttons, or even motion sensors. But with the N1 implant, the interface is direct. The brain itself becomes the controller.

Participants in the Convoy Study will have the ability to direct the robot's movements with nothing more than their thoughts, making this not just a technological breakthrough, but a profound step toward creating a seamless connection between human cognition and machine action.

For people with neurological impairments, this study has the potential to be life-changing. Imagine a person who has lost the ability to move their arms or legs due to paralysis being able to control a fully functional robotic arm with their thoughts. This capability isn't some distant, far-off vision—it's actively being tested. The Convoy Study's success could lead to a future where people can regain mobility or perform complex tasks without the need for traditional prosthetics or assistive devices. With the right integration of brain and machine, the limitations of the human body could be transcended, allowing individuals to engage with the world in new, empowering ways.

Moreover, the Convoy Study holds immense significance for the broader field of robotics and artificial intelligence. If Neuralink can successfully demonstrate that the brain can control a humanoid robot like Optimus with the same precision as a biological limb, it will open the door to an entirely new era of robotics. One where AI and human control are seamlessly integrated, allowing robots to perform complex tasks with a human touch. This could revolutionize industries ranging from healthcare, where robots might assist in surgeries or patient care, to manufacturing, where humanoid robots could perform tasks that require both dexterity and adaptability.

As Neuralink continues its work with the Convoy Study, it is clear that the implications stretch far beyond restoring mobility for individuals with disabilities. It signals the dawn of a new age of human enhancement, where the mind's power can drive machines to perform tasks previously confined to human bodies. The study represents not just a scientific experiment but a roadmap to a future where the line between human and machine is increasingly blurred. The results of the Convoy Study may very well be the key that unlocks the future of brain-machine integration, where we no longer just use technology, but become one with it.

Chapter 2: Tesla Optimus: The Future of Humanoid Robots

Tesla Optimus is more than just a robot; it's a vision of the future of robotics, designed to seamlessly integrate into human environments and replicate human biomechanics with unparalleled precision. Announced by Elon Musk in 2021, Optimus was conceptualized to be a humanoid robot capable of performing a wide range of tasks that require the kind of dexterity and flexibility that humans take for granted. Unlike traditional robots that are often limited by rigid programming or mechanical limitations, Optimus was designed with human-like movement in mind, offering a level of precision and adaptability that sets it apart from anything we've seen before.

At the heart of Optimus' design is its ability to move and behave like a human, with a focus on replicating natural human biomechanics. Tesla's engineers studied the intricacies of human movement, from walking to handling objects, to ensure that Optimus could perform these tasks with the same fluidity and grace. The goal isn't just to create a robot that can move but to create one that can do so in a way that feels organic, much like a human would. Whether it's picking up groceries, performing simple household tasks, or assisting in industrial applications, Tesla Optimus is built to integrate seamlessly into environments where human-like movement is essential.

One of the standout features of Tesla Optimus is its advanced hand design, which boasts an impressive 22 degrees of freedom. In comparison, the human hand has 27 degrees of freedom, making it one of the most dexterous parts of the human body. While most robotic hands are limited to just six or seven degrees of freedom, Optimus' hands can move with incredible precision and flexibility, allowing it to manipulate objects with finesse and perform tasks that require both strength and delicacy. This enhanced dexterity is crucial for tasks like assembling intricate parts, handling fragile objects, or even performing complex surgeries in the future.

The design of Optimus' hands was a key area of focus for Tesla engineers, as they aimed to create a robot that could replicate the complex movements of the human hand. To achieve this, they incorporated a wide range of joints, actuators, and sensors that allow the robot to move its fingers, wrists, and thumbs in ways that mimic the human grasp. The robot can pick up small items, grip larger objects, and adjust its hold with exceptional accuracy. This level of dexterity is one of the reasons why Tesla Optimus is considered one of the most advanced humanoid robots in the world.

Additionally, Tesla's engineers enhanced the robot's overall flexibility by incorporating more degrees of movement in its form and wrist. These upgrades have doubled the flexibility of earlier models, providing Optimus with even greater mobility and the ability to handle tasks that require both strength and precision. Whether it's lifting a heavy object or delicately adjusting a tiny screw, Optimus' hands and wrists are equipped to tackle the full range of motions a human might need.

But the robot's dexterity doesn't stop at its hands. Optimus is designed with a complete range of movement that allows it to walk, balance, and perform tasks with human-like coordination. Unlike traditional robots, which often struggle with maintaining balance or making coordinated movements, Optimus uses sophisticated algorithms and sensors to move fluidly, adjusting its posture and gait to mimic human walking patterns. This allows the robot to navigate complex environments, such as homes or workplaces, where uneven terrain or obstacles might be present.

The flexibility and precision built into Tesla Optimus make it a remarkable piece of engineering, not only for its potential applications in personal assistance and manufacturing but also for the broader field of robotics. Its ability to mimic human movement so closely is a significant leap forward in creating robots that can work alongside humans in everyday settings, interacting with their surroundings and performing tasks that require a high degree of manual skill.

As the development of Tesla Optimus continues, its key features of precision, flexibility, and dexterity will likely become even more refined. This progress will open up new possibilities in fields ranging from healthcare to manufacturing, where human-like robots can assist with both complex tasks and everyday chores. The future of robotics, as envisioned by Tesla, is one where robots and humans collaborate seamlessly, with Optimus leading the way.

Tesla's journey to develop the Optimus humanoid robot has been nothing short of ambitious, with each step marked by significant technological milestones. The development timeline of Optimus showcases the progress of Tesla's engineers and AI experts, pushing the boundaries of what robots can do and how they can interact with the real world.

Since its unveiling in 2021, Optimus has undergone rapid refinements, with each phase bringing it closer to becoming a fully functional robot capable of performing human-like tasks across various industries.

The initial announcement of Tesla Optimus in August 2021 was a bold vision for the future of robotics. Elon Musk introduced the project at Tesla's AI Day, outlining the goals of creating a humanoid robot that could help with "mundane and repetitive tasks" in daily life. At that point, the robot was still in its conceptual phase, with the team focusing on developing the underlying technology that would make it possible to build a robot with human-like movement and dexterity. Musk's vision was clear: Optimus would be designed to handle tasks in environments like homes, warehouses, and even factories—places where manual labor and repetitive tasks dominate.

By 2022, Tesla was able to reveal a working prototype of Optimus. While the robot's movements were initially stiff and limited, the unveiling marked a huge step forward in the development process. During the presentation, the prototype demonstrated its ability to walk, wave, and perform basic movements, offering a tangible example of what had previously been a vision. The early prototype wasn't designed for complex tasks, but it was a proof of concept that showed Tesla's engineers had successfully created the foundational aspects of Optimus. The demonstration emphasized Tesla's commitment to developing a robot that could integrate seamlessly into human environments, with a focus on biomechanics and dexterity.

In the following year, 2023, Tesla made significant advancements to improve Optimus' range of motion, stability, and overall functionality. The robot's ability to handle more intricate tasks, such as picking up objects and performing delicate movements, became one of the focal points of its development. Tesla engineers worked on refining the robot's artificial intelligence systems, which help it learn from its surroundings and adapt to new tasks. One of the key breakthroughs was the improvement of its hands, which now featured 22 degrees of freedom, a level of dexterity far surpassing that of most robots. This allowed Optimus to manipulate objects with greater precision, handling tasks that would have been unthinkable for earlier robotic models.

Tesla's focus on versatility has been a driving force behind Optimus' development. The company has emphasized creating a robot that can be used in a wide range of real-world applications, particularly in industries that could benefit from automation. Healthcare, logistics, and industrial sectors have all been identified as prime candidates for Optimus integration. In healthcare, the robot could be used for physical therapy, assisting patients with mobility issues, or performing routine tasks such as moving equipment and supplies. With its dexterous hands, Optimus could even assist in surgical environments, providing precise movements for delicate procedures.

In logistics and industrial settings, Optimus could drastically improve efficiency by performing tasks that would typically require human labor. Whether it's moving heavy materials in a warehouse, assisting in assembly lines, or managing inventory, the robot's versatility makes it an ideal candidate for streamlining operations and reducing the strain on human workers. Optimus' ability to handle repetitive, labor-intensive tasks could free up human workers to focus on more complex and creative problem-solving roles, thus improving productivity and worker safety.

Tesla's focus on durability has been equally important in ensuring that Optimus can handle the demands of these real-world applications. The robot's design incorporates high-quality materials and robust components, allowing it to withstand the wear and tear of daily use in diverse environments. Optimus is built to be reliable, with a focus on minimizing maintenance requirements and ensuring longevity in industries that rely on consistent performance. This durability is especially critical in healthcare and industrial environments, where downtime can be costly.

Tesla's engineers have also worked to improve Optimus' cognitive abilities, allowing it to understand and adapt to its surroundings in real time. Equipped with a suite of advanced sensors, cameras, and AI algorithms, Optimus can perceive its environment and make decisions based on what it detects. Whether it's navigating through a crowded hospital corridor or managing the logistics of a bustling warehouse, Optimus can respond to obstacles, recognize patterns, and adjust its behavior accordingly. This autonomy is essential for making the robot functional in unpredictable, dynamic environments where human-like decision-making is required.

Looking ahead, Tesla's roadmap for Optimus is filled with exciting possibilities. The company's focus is on refining the robot's design and AI capabilities to ensure that it can perform complex tasks in real-world scenarios. As the robot's development progresses, Tesla aims to introduce Optimus into everyday use, where it can assist in homes, businesses, and industries across the globe. While there is still a long way to go before Optimus becomes a ubiquitous part of daily life, the milestones Tesla has already achieved suggest that the future of robotics, powered by AI and human-like precision, is closer than ever before. Optimus may soon become a trusted partner in countless industries, helping people live more efficiently, safely, and comfortably.

Chapter 3: How Neuralink and Optimus Work Together

The Neuralink N1 implant is a groundbreaking technology that bridges the gap between the human brain and machines, allowing individuals to control devices and robotic limbs using only their thoughts. At its core, the N1 implant is a brain-computer interface (BCI) designed to decode neural signals in real time. These signals are the electrical impulses that our brains use to send commands to our muscles and control movements. By tapping into these signals, Neuralink has developed a system that can interpret a person's intentions and translate them into actions, whether it's moving a cursor on a screen, controlling a robotic arm, or interacting with other types of technology.

The implant itself is a small, wireless device, about the size of a coin, that is surgically placed in the brain. It sits in the region responsible for motor control, the part of the brain that sends signals to the body's muscles. The device is equipped with thousands of tiny, flexible electrodes that can record neural activity with unprecedented precision. These electrodes detect the electrical signals produced by neurons as they fire, sending this information to an external computer that decodes it into meaningful commands.

Once implanted, the N1 device creates a direct communication channel between the brain and the machine. The brain's signals are transmitted wirelessly to a nearby receiver, which processes the data and sends it to the robotic or digital system being controlled. For example, when someone thinks about moving their hand or leg, the N1 implant picks up these brain signals, interprets them, and translates them into corresponding movements in a robotic limb or digital interface. The result is a seamless connection between thought and action, with little delay or interference.

The N1 implant's ability to decode brain signals and control machines has significant implications, particularly for individuals with disabilities. One of the most compelling real-world examples of this technology's potential is its use with quadriplegic patients. These individuals, who have lost the ability to move their limbs due to spinal cord injuries, often face severe limitations in interacting with the world. However, with the N1 implant, some quadriplegic patients have been able to regain a level of control over their environment that was previously unimaginable.

In clinical trials, participants who were unable to use their hands or arms have demonstrated the ability to control digital devices such as smartphones, computers, and video game consoles using only their thoughts. In one notable case, a 29-year-old quadriplegic named Nolan Arbau was able to play chess and video games without the use of his hands, simply by thinking about his moves. The N1 implant allowed him to control the action on the screen with remarkable accuracy, demonstrating that the brain could still function as a natural interface, even when the body's motor functions were impaired.

These examples are just the beginning. The N1 implant has the potential to help patients regain a wide range of abilities. As the technology continues to evolve, the next logical step is to extend its application to controlling physical prosthetics or robotic limbs. This would enable individuals with paralysis to move robotic arms, legs, or even full-body prosthetics with the same level of control they would have over their biological limbs. For people who have lost their ability to move due to injury or illness, this could be life-changing, offering a new form of mobility and independence.

For example, imagine a person with quadriplegia being able to move a robotic arm or leg with their mind to perform tasks like drinking from a cup, opening a door, or even shaking hands. The N1 implant's potential to restore lost function is nothing short of revolutionary. By bypassing the damaged spinal cord and connecting directly to the brain, it offers a new way to control movement, helping people with severe disabilities regain independence and autonomy.

The real-world implications of the N1 implant go beyond just helping individuals with physical disabilities. The technology could be applied to a wide range of areas, from robotics and prosthetics to virtual reality and augmented reality. The ability to control digital systems, robots, or other technology with thought alone opens up new possibilities for enhancing human capabilities in virtually any field. It could transform industries like healthcare, manufacturing, education, and entertainment, creating new ways for people to interact with machines and their environments.

As Neuralink continues to refine the N1 implant and expand its applications, the dream of a world where people can control technology with their minds is becoming more of a reality. The technology is still in its early stages, but the possibilities it offers are already beginning to transform the way we think about the relationship between humans and machines. The ability to control robotic limbs with thoughts alone represents not just a breakthrough in medical technology but also a glimpse into a future where the boundaries of human potential are pushed further than ever before.

In 2023, one of the most remarkable demonstrations of the Neuralink N1 implant's potential came with the story of a 29-year-old participant named Nolan Arbau. A quadriplegic, Nolan had been unable to move his limbs following a severe spinal cord injury. His life, once filled with the ability to perform simple tasks like typing, eating, and playing sports, had been drastically altered. But with the advent of the Neuralink N1 implant, Nolan's experience of disability was set to change in a way that defied the limits of traditional medicine and rehabilitation.

Nolan's journey began when he enrolled in one of Neuralink's clinical trials, designed to test the N1 implant's ability to help people with severe spinal cord injuries regain control over digital devices. After the implant was surgically placed in his brain, it was connected to a computer that could read his brain's motor signals. The implant's electrodes were strategically positioned in areas of Nolan's brain responsible for hand and arm movements, so it could pick up his thoughts as he attempted to move his limbs. However, the breakthrough wasn't limited to regaining limb movement—it extended to the power of thought itself.

What followed was nothing short of extraordinary. Using the N1 implant, Nolan was able to control a digital interface with his mind. He played chess, moved pieces on a screen, and even engaged in video games—tasks that most of us take for granted, but which were previously inaccessible to Nolan due to his paralysis. The result was a seamless translation of his brain's signals into actions, demonstrating that thought could once again interact with technology, just as it would with a functioning body. There was no need for external devices like joysticks, mice, or keyboards—his mind alone became the controller.

What was particularly groundbreaking about Nolan's experience was not just the ability to play games or manipulate a computer but the demonstration of the brain's adaptability and the N1 implant's capacity to decode complex neural signals. Despite the physical barriers caused by his injury, Nolan's brain still retained the ability to produce the neural commands needed for movement. The N1 implant was simply tapping into those signals, bypassing the damaged spinal cord, and translating them into a digital action. The technology restored Nolan's ability to interact with the digital world, not through physical movement but through the direct interface of his brain.

This case study is a powerful example of how brain-computer interfaces like Neuralink's N1 implant are making strides in not only restoring lost functions but also redefining the relationship between humans and machines. The technology doesn't merely act as a tool but as an extension of the body, allowing the mind to control systems directly and fluidly.

The collaboration between Neuralink and Tesla is pushing the boundaries of human potential even further. Tesla's cutting-edge robotics, particularly the Tesla Optimus humanoid robot, is a key partner in this vision. While the N1 implant is focusing on interpreting brain signals to control digital devices and basic prosthetics, the next phase is about moving those same signals into the realm of complex robotics.

Tesla's Optimus humanoid robot is designed to be a lifelike, autonomous machine capable of performing everyday tasks with human-like precision and flexibility. By combining the advancements of Neuralink's N1 implant with the capabilities of Optimus, we are looking at the possibility of individuals with disabilities controlling robotic limbs—or even entire robotic bodies—with the same fluidity and naturalness as biological limbs. This collaboration is not just about overcoming physical impairments; it's about enhancing human capabilities and pushing what is possible for all people, regardless of their physical conditions.

For Nolan Arbau, the potential future outcome could be even more profound. While his ability to play chess and interact with a digital screen is an amazing feat, the next logical step would be to use the N1 implant to control a Tesla Optimus robot or a prosthetic limb in a similarly seamless way. Imagine him—after years of paralysis—being able to move a robotic arm, walk, or perform tasks that require fine motor control, simply by thinking about them. This technology, when combined with Tesla's highly advanced robotics, could provide people with disabilities a level of independence never before possible.

Furthermore, the implications for individuals who are not disabled are equally powerful. By fusing brain-computer interfaces with advanced robotics, the potential exists for anyone to enhance their own physical capabilities. Whether it's controlling a robotic assistant to perform household chores, interacting with virtual environments, or even enhancing physical strength or dexterity, the possibilities are endless.

The convergence of Neuralink's brain-computer interfaces and Tesla's robotics is leading us into a new era where the line between human thought and action is becoming increasingly blurred. What was once science fiction is now on the cusp of becoming a reality—an age where people can seamlessly control machines, not just through their hands or voice, but through the power of their minds. This collaboration is reshaping the future of human ability, offering a world where the limitations of the body no longer define what's possible.

The prospect of controlling entire bodies or complex robotic systems with nothing but thought is a vision that once seemed far-fetched, relegated to the realms of science fiction. But with the ongoing advancements in brain-computer interfaces (BCIs) and robotics, particularly with innovations like Neuralink's N1 implant and Tesla's Optimus humanoid robot, this futuristic concept is beginning to materialize in the present day.

At the heart of this breakthrough is the N1 implant, a device that allows the human brain to directly communicate with machines. Neuralink's technology is grounded in the principle of translating neural activity—the electrical signals generated by neurons in the brain—into actionable commands. With the N1 implant, these signals are captured by the implant's electrodes and transmitted wirelessly to external systems, allowing individuals to control devices, computers, and even robotic limbs with thought alone. Initially, the focus has been on helping people with severe disabilities regain some level of control over their environment. However, the real promise lies in extending this capability to more complex systems, such as full-body exoskeletons or sophisticated humanoid robots.

Imagine a future where a person, paralyzed from the neck down due to a spinal cord injury, can control a fully functional robotic body with their mind. This would not only allow them to perform simple tasks like reaching for a cup or walking across a room, but also enable more complex actions—bending, stretching, lifting, or even running—all driven by the user's thoughts. The beauty of this concept lies in its seamless integration between the brain and the robotic systems. There would be no need for physical input devices like joysticks or touchscreens. Instead, the brain would directly instruct the robot's movements, mimicking natural actions with ease.

The potential for controlling entire robotic bodies or systems could redefine how we think about mobility and human capability. For those who have lost the ability to move their own bodies due to paralysis or injury, a robotic body controlled entirely by thought offers the possibility of regaining not just independence, but also a sense of normalcy. But the dream doesn't end with just regaining function—it extends to enhancing human potential. Even individuals without physical disabilities could use BCIs to control robotic systems that augment their own bodies, whether through advanced prosthetics or full-body exoskeletons.

Tesla's Optimus robot is a prime example of how these technologies could evolve. The Optimus robot, with its 22 degrees of freedom in its hands and advanced dexterity, is already capable of performing tasks like lifting, carrying, and interacting with objects in ways that human hands do. The idea of linking the N1 implant to Optimus, or similar robotic systems, could allow users to control not just a single robotic limb but a whole humanoid robot, effectively acting as an extension of their own body. For example, a person could think about picking up a box or turning a knob, and the robot would do so with the precision of a human hand. But unlike a human body, the robot would be free from fatigue, pain, or physical limitations, offering limitless potential for human performance.

One of the most exciting aspects of this fusion of mind and machine is its potential to restore, and even enhance, human abilities. For instance, individuals with paralysis could experience not only a return of their motor functions but a heightened sense of independence, performing tasks and engaging with the world around them without relying on others. Furthermore, by controlling robotic systems, a person could interact with their environment in ways that were previously impossible—lifting heavy objects, performing delicate surgical tasks, or operating machinery that requires precise, repetitive movements.

In industrial and healthcare settings, this could mean that robotic workers could be operated by people with disabilities, performing complex tasks with ease and efficiency. For instance, a quadriplegic could operate a robotic exoskeleton to assist with physical therapy or even use it to work in a factory or warehouse, performing tasks that would typically require human labor. The implications are profound: instead of being limited by their own bodies, people could use their minds to operate sophisticated machines, enabling them to take on jobs, hobbies, and roles in society that they would otherwise be unable to.

The promise of controlling entire robotic systems with thought alone also extends to the realm of virtual and augmented reality. With the ability to manipulate robotic bodies or systems seamlessly, people could engage in virtual worlds or remote environments with a level of interactivity that feels completely natural. Imagine being able to navigate a complex virtual environment with your mind, controlling an avatar or robot in real-time without the need for controllers or external devices. This kind of freedom would not only revolutionize entertainment and gaming but also extend to remote work and training, enabling users to interact with complex systems or environments from anywhere in the world, all controlled by thought.

In the long run, the fusion of brain-computer interfaces and robotics will likely reshape society in ways we can't yet fully predict. With the ability to control entire bodies or robotic systems through thought, the line between what is considered "human" and "machine" will blur, opening up new avenues for human augmentation and creating opportunities for people to live fuller, more independent lives. This future is not just about overcoming physical limitations—it's about enhancing human capabilities to their fullest extent, unlocking potential that was previously confined to the realm of imagination.

As the N1 implant and Tesla Optimus robot continue to evolve, we are witnessing the dawn of a new era in human-machine interaction, one where the only limit is the imagination. Whether for medical rehabilitation, industrial automation, or personal enhancement, the ability to control robotic systems with thought alone is set to redefine human capabilities, unlocking a future where the mind becomes the ultimate tool for shaping the world around us.

Chapter 4: Robotics and Human Mobility: Restoring and Enhancing Abilities

The integration of the Neuralink N1 implant with robotic limbs marks a significant step forward in the quest to restore mobility to individuals who have lost the ability to move their limbs due to spinal cord injuries, neurological diseases, or other debilitating conditions. The N1 implant itself acts as the neural interface, allowing the brain to directly communicate with external devices, bypassing damaged spinal cords or compromised motor functions. When used in conjunction with robotic limbs, the N1 implant creates a seamless link between the brain's signals and the mechanical movements of the robotic limbs.

The process begins with the implantation of the N1 device, which consists of thousands of ultra-thin, flexible electrodes that are placed in specific areas of the brain responsible for motor control. These electrodes pick up electrical signals generated by neurons as they send commands to the muscles.

For individuals with paralysis, these neural signals may no longer have the ability to reach the body's muscles due to spinal cord injuries or diseases like ALS. However, the N1 implant can interpret these signals and translate them into commands that robotic limbs can understand and execute.

Once the N1 implant is in place and fully functional, individuals can use their thoughts to control robotic prosthetics or assistive devices in ways that mirror natural limb movements. For example, a person could think about moving their hand to grab a cup, and the robotic arm, equipped with sensors and actuators, would move in response to those mental commands. The brain does not need to physically generate the movement itself—instead, it simply sends the "thought" or intent to the implant, which decodes the signal and directs the robotic limb to act accordingly.

The beauty of this technology lies in its naturalness and fluidity. With proper calibration and fine-tuning of the implant's decoding abilities, the movements of the robotic limbs can mimic the user's natural gestures with incredible precision. It's not just about moving a robotic arm; it's about recreating the subtleties of human motion. The system can enable complex actions such as grasping objects, lifting, manipulating delicate tools, or even performing tasks like typing on a keyboard. The N1 implant's ability to interpret the brain's intentions means that the user can control these movements intuitively, without needing to learn how to operate the robotic limb through complex gestures or external controllers.

In the future, the potential of this technology could extend far beyond just prosthetics. Neuralink and Tesla's collaboration holds the promise of developing full-body robotic systems that can be controlled with thought alone. This could lead to the creation of exoskeletons or full-body prosthetics that replicate natural human movement. Imagine a person with paralysis being able to walk again by controlling a robotic exoskeleton that mimics the movement of their own legs. Using only their thoughts, they could take steps, climb stairs, or navigate obstacles with the same fluidity as if they were using their own body.

Tesla's Optimus humanoid robot represents a key part of this vision. While current prosthetic limbs are relatively limited in their ability to replicate the full range of human movement, Optimus is designed to have the kind of flexibility, dexterity, and mobility that could one day replace traditional prosthetics and assistive devices. With its advanced capabilities, including 22 degrees of freedom in its hands and the ability to perform delicate tasks like catching a tennis ball, Optimus offers a level of precision that current prosthetics simply can't match. In the future, Optimus could be equipped with the N1 implant, enabling users to control the robot with their minds, effectively extending the human body beyond its natural limitations.

For individuals with disabilities, this could be revolutionary. Prosthetics have historically been clunky, difficult to control, and limited in their function. By integrating the N1 implant with a sophisticated humanoid robot like Tesla Optimus, the possibilities become far more expansive. Instead of simply using a robotic arm to perform basic tasks, individuals could control a fully functional humanoid robot that could walk, talk, and interact with the world in ways that mirror human behavior. This type of robot could replace not only traditional prosthetics but also assistive devices like wheelchairs, giving users a newfound sense of freedom and independence.

What sets this combination of Neuralink and Tesla apart from traditional assistive technologies is its precision and naturalness. Current prosthetics and assistive devices often require extensive training to operate effectively, and they rarely offer the fluidity or responsiveness of natural human movement. In contrast, the Neuralink system promises to decode neural signals with such precision that the user can control the robotic limb or full-body exoskeleton as naturally as they would their own body. This intuitive connection between the mind and the machine is what makes this technology so transformative.

In addition to restoring mobility, this integration could also improve the quality of life for people with disabilities by reducing the physical and emotional toll that traditional assistive devices often impose. For example, a person who has relied on a wheelchair for years might suddenly find themselves able to stand and walk with the help of a robotic exoskeleton. The emotional impact of regaining mobility cannot be overstated, as it restores a level of dignity and independence that has been lost.

The implications of this technology extend beyond just prosthetics. Imagine a world where people can enhance their physical capabilities by controlling robotic systems with thought alone. This could open up a new era in industries such as healthcare, manufacturing, and logistics, where robotic limbs and exoskeletons can be controlled directly by the human brain to perform tasks that are otherwise impossible or unsafe for human workers. In the medical field, it could enable surgeons to operate with even more precision and dexterity, using robotic arms controlled by their thoughts.

As Neuralink and Tesla continue to refine their respective technologies, the dream of using thought to control complex robotic systems is becoming a reality, and the possibilities for people with disabilities are more exciting than ever. Whether it's regaining basic mobility, performing intricate tasks with robotic limbs, or even using exoskeletons to navigate the world, the fusion of brain-computer interfaces and advanced robotics is set to redefine what it means to be human in the 21st century.

The future possibilities for the integration of Neuralink and Tesla Optimus go far beyond the restoration of basic functions like mobility. As the technology advances, it holds the potential to significantly enhance human capabilities in ways that we could only imagine a decade ago. Imagine a world where the human mind controls robotic limbs or full-body exoskeletons with precision and power, enabling individuals to perform tasks that were once impossible or incredibly difficult.

One of the most exciting prospects is the ability to lift and manipulate heavy objects with ease. Individuals with physical disabilities, such as those with paralysis or amputations, could control robotic arms or exoskeletons that possess the strength and dexterity to perform tasks that typically require significant human effort. Lifting heavy boxes, moving industrial materials, or even performing demanding physical labor could be done by anyone, regardless of their physical limitations. This opens up vast possibilities in industries such as logistics, construction, manufacturing, and warehousing, where manual labor and physical strength are paramount. Instead of relying on a team of workers to lift and move heavy items, a person using a robotic exoskeleton could complete the task independently, without breaking a sweat.

The precision and dexterity of Tesla Optimus, with its 22 degrees of freedom in the hands and ability to perform delicate tasks like catching a tennis ball, could allow individuals to perform intricate, high-skill jobs with ease. Surgeons could use robotic systems controlled by their thoughts, conducting surgeries with the precision of a machine while benefiting from the natural human intuition that guides their actions. In manufacturing, workers could operate robotic systems that handle delicate machinery or perform tasks like welding, assembly, and inspection with far more accuracy than a human could achieve. The robotic system could perform repetitive tasks without fatigue, which would increase productivity while reducing the likelihood of human error. This could drastically transform industries that rely on high-precision tasks, such as electronics manufacturing, aerospace, and medical device production.

In addition to enhancing manual labor, the combination of the N1 implant and Tesla Optimus could fundamentally alter the way we interact with the physical world. Instead of relying on traditional tools, machines, or even complex software, a person could simply think about the task at hand, and the robotic system would execute it flawlessly. Whether it's adjusting a machine's settings, assembling a car, or creating detailed artwork, the possibilities are virtually endless.

The most profound impact, however, may be seen in healthcare, particularly in how individuals with disabilities interact with the world around them. For many people living with severe physical impairments, daily tasks that others take for granted can be a constant struggle. Tasks like dressing, eating, or even moving around can require the assistance of caregivers or family members. But with the advent of robotic systems that can be controlled with the mind, the reliance on caregivers could be significantly reduced, offering people more independence and autonomy. A person with quadriplegia could operate a robotic exoskeleton to walk, move objects, or even perform personal hygiene tasks—all with nothing more than their thoughts.

This reduction in dependence on caregivers would not only provide individuals with disabilities greater freedom but also alleviate some of the emotional and financial burdens faced by caregivers. Family members and professional caregivers would no longer need to be present for every moment of a disabled person's life, freeing up time for both the caregiver and the individual receiving care. The impact of this autonomy could be profound, allowing people with disabilities to live more fulfilling, independent lives and pursue careers, hobbies, or social activities that were previously out of reach.

As we look further into the future, these advancements could also lead to a reimagining of what it means to be human. The integration of brain-computer interfaces with robotics could pave the way for enhanced human abilities—taking our natural talents and augmenting them with the precision and strength of machines. Some may even see this as the dawn of a new era where physical limitations are no longer a defining factor of one's abilities.

For example, someone who has lost the ability to use their legs could regain the ability to walk, not just with a traditional prosthetic but with a full-body exoskeleton that mimics natural movement. This would not just restore mobility; it would amplify it. A person could potentially run, jump, or engage in athletic activities once again. Even individuals without disabilities could use robotic enhancements to push beyond their natural limits—lifting heavier weights, achieving more precise movements, or performing tasks at a faster pace. In sports, for example, the combination of human strength and robotic precision could create a new class of athletes, capable of feats that were previously unimaginable.

Ultimately, the combination of Neuralink's brain-computer interface and Tesla's robotics could mark the beginning of a new chapter in human evolution. Instead of accepting the limitations of our bodies, we could embrace a future where those boundaries are redefined, offering everyone—regardless of their physical abilities—the chance to transcend their human limitations and become something greater. The promise of this technology is not just to restore what was lost, but to enhance what is possible, creating a future where human potential knows no bounds.

Chapter 5: Neuralink and Tesla: Shaping the Future of Human Enhancement

Elon Musk's vision for Neuralink goes far beyond its initial mission of restoring mobility and helping people with neurological impairments. For Musk, the fusion of human minds with advanced technology is not merely about rehabilitation—it's about pushing the boundaries of human potential. His vision hints at a future where humans are no longer constrained by their biological bodies. Instead, by integrating brain-computer interfaces (BCIs) like Neuralink's N1 implant, people could enhance their physical and cognitive abilities to extraordinary levels, even creating what Musk has referred to as "cybernetic superpowers."

The idea of "cybernetic superpowers" centers around the notion that by connecting the human brain to sophisticated machines, we could augment not just our physical capabilities but also our mental faculties. Consider the power to control robotic limbs, entire exoskeletons, or even robotic systems through thought alone, with no need for physical gestures or complex devices. The technology could extend to enhancing muscle strength, endurance, and dexterity to levels that far exceed what the human body can naturally achieve. With Neuralink, the brain would be able to communicate directly with these enhanced systems, enabling seamless interaction between mind and machine.

Beyond mere restoration of function, the N1 implant could elevate human abilities in ways that were once the realm of science fiction. Imagine being able to lift objects that would normally be far too heavy to manage, or performing intricate tasks with far more precision than human hands are capable of. For example, an individual could think about performing a complex operation or delicate task, and the robotic system would carry out these actions with the fine motor control and dexterity of a highly advanced machine. This kind of enhancement could revolutionize industries like surgery, precision manufacturing, and construction, where the need for delicate handling and heavy lifting often limits human performance.

But the applications of Neuralink technology don't end with physical augmentation. Cognitive abilities could also be enhanced in profound ways. One of the most exciting possibilities for Musk is the potential to merge human intelligence with machine intelligence, allowing people to access vast amounts of information almost instantly, process complex data at a much faster rate, and even communicate directly with machines or other humans via thought. Such a leap could radically alter fields like education, research, and even creativity. Imagine being able to download new skills directly into your mind, or interface with a computer without needing to type or click. Neuralink's advanced BCI could potentially grant users the ability to expand their cognitive horizons, empowering them to learn, think, and solve problems in ways that would have been unimaginable before.

Elon Musk has often discussed his long-term vision of humans becoming "symbiotic" with AI, creating a hybrid between the biological and the technological. The goal is not just to overcome the limitations of human biology but to create a new form of human potential that harnesses the power of both mind and machine. The idea is that by integrating neural implants like N1 into our daily lives, we could experience an unprecedented expansion of what it means to be human—where mental and physical barriers are no longer limitations, and the possibilities are as vast as human imagination.

This vision isn't just a far-off dream. The groundwork is already being laid with the ongoing trials and developments in Neuralink technology. Each step brings us closer to a world where individuals can regain lost abilities, enhance their physical performance, and unlock their cognitive potential. As these technologies evolve, they will likely become a common part of life for millions, not just for those with disabilities, but for anyone seeking to expand the limits of their abilities.

For Musk, this is just the beginning. The concept of "cybernetic superpowers" is rooted in the belief that, as technology becomes increasingly advanced, it will no longer be enough for humans to merely live alongside machines. Instead, we will merge with them, becoming something greater than we ever thought possible—both biologically human and technologically enhanced. Whether it's lifting heavy objects, performing precise surgeries, or thinking at speeds that rival the fastest computers, Neuralink is poised to play a central role in unlocking a new era of human evolution, where the boundaries between man and machine blur in ways that offer limitless potential.

The future of AI-powered robotics integrated with the human brain holds immense potential across a wide range of industries. As Neuralink's brain-computer interface (BCI) technology and

Tesla's robotics continue to evolve, we may soon see a dramatic transformation in how industries operate, with humans and machines working in harmony to achieve unprecedented levels of efficiency, precision, and strength.

In construction, for example, the integration of AI-powered robotics with the human brain could completely revolutionize the way buildings are designed and constructed. Traditionally, construction relies heavily on manual labor and machines that require human input. However, with the advent of robotic systems controlled by thought alone, workers could use their minds to control powerful exoskeletons or advanced robotic arms to perform labor-intensive tasks like lifting heavy materials, assembling complex structures, or even fine-tuning intricate components with remarkable precision. This could lead to faster construction timelines, reduced labor costs, and safer working conditions by eliminating the need for workers to perform physically demanding tasks that often put them at risk of injury. Additionally, AI-powered machines could enhance the ability to perform complex tasks with precision that would be challenging for humans alone, such as fitting parts with millimeter accuracy or performing welding tasks in hazardous environments.

In logistics, the integration of robotics and BCIs could completely redefine supply chain management and warehouse operations. Robots powered by AI and controlled by human thought could autonomously navigate warehouses, manage inventory, and deliver goods with exceptional speed and accuracy. For example, a warehouse worker could mentally direct a robotic system to fetch an item from a high shelf, sort packages, or load and unload trucks, all while avoiding obstacles and adapting to real-time changes in the environment. This level of automation, driven by human intuition and AI, could drastically increase productivity and reduce errors in logistics operations, leading to faster delivery times and more efficient management of resources.

The potential impact on athletics and physical training is another fascinating area where the integration of AI-powered robotics and BCIs could lead to new breakthroughs. Athletes could utilize exoskeletons or robotic limbs that enhance their strength, endurance, and precision while simultaneously allowing them to push their bodies to new limits. The technology could also be used to monitor and optimize athletic performance in real time, allowing coaches and athletes to gain detailed insights into the biomechanics of movement, muscle strain, and recovery. Imagine an athlete wearing a full-body exoskeleton that not only provides physical support but also enhances their movements with greater power and agility. This could open up new possibilities for athletes, enabling them to reach peak performance without the limitations of human biology.

For individuals with disabilities, this technology could mean a complete transformation of what's possible. Robotic systems that are controlled with the mind could allow people to perform everyday activities with ease, regaining independence and enhancing their quality of life. Whether it's walking, lifting objects, or interacting with the environment in ways previously thought impossible, the ability to control AI-powered robotics with thought alone would open up a new world of possibilities for people with mobility impairments.

One of the key challenges to making this revolutionary technology accessible to the wider public lies in mass production and scaling. Neuralink and Tesla have made significant strides in developing their technologies, but for them to have a widespread impact, they must be produced at a scale that makes them affordable and practical for a broad range of industries and individuals. Mass production is crucial for driving down the cost of brain-computer interface implants, robotic systems, and exoskeletons. As these technologies become more refined and efficient, the price of the necessary hardware and implants will likely decrease, making them accessible to more people and industries. Tesla's expertise in manufacturing and mass production, along with Neuralink's advancements in neural interface technology, places them in a prime position to scale these innovations in a way that could bring them into the mainstream.

Tesla's approach to mass production is rooted in the same principles that have made their electric vehicles more affordable and widely adopted over the past decade. By optimizing manufacturing processes, reducing the cost of components, and leveraging automation, they are able to produce cutting-edge technologies at a lower cost, making them accessible to a larger market. If Tesla can apply these same principles to the production of humanoid robots like Optimus and integrate them with Neuralink's brain-computer interface technology, the result could be a breakthrough in making advanced robotics and brain enhancements available to the general public.

Scaling the technology will also require significant advances in medical implantation procedures. Neuralink is already working on making the implantation process as minimally invasive and safe as possible, with the goal of reducing the risks and costs associated with surgery. As more people undergo these procedures and as the technology becomes more refined, the entire process could become more efficient, lowering both the initial cost and the long-term maintenance of these systems.

For businesses, scaling these technologies would also mean adopting a more seamless integration of human and machine workers. Companies in industries like construction, logistics, and healthcare could benefit from robotic systems controlled by employees' thoughts, improving efficiency, reducing human error, and lowering physical strain on workers. This could also lead to safer work environments, particularly in high-risk fields where human intervention is often limited by the hazards of the job.

Ultimately, the goal of Neuralink and Tesla is not just to enhance the abilities of individuals but to transform entire industries, creating a future where AI-powered robotics and human brains work together as an integrated team. Through mass production, strategic scaling, and continuous innovation, these technologies have the potential to unlock capabilities that were once confined to the realm of science fiction, creating a new paradigm in which humans and machines are not separate but symbiotic, advancing both human potential and the world we live in.

Chapter 6: Ethical, Social, and Security Implications

As with any groundbreaking technology, the widespread use of Neuralink and Tesla Optimus raises complex ethical questions that society must confront. These technologies, which blur the lines between human biology and advanced robotics, have the potential to transform human existence in profound ways. However, with these advancements come critical considerations about autonomy, privacy, consent, and the very nature of what it means to be human.

One of the most pressing ethical concerns is the issue of identity. As Neuralink and Tesla Optimus make it possible to control robotic limbs or even full-body exoskeletons with the mind, the distinction between human and machine becomes increasingly difficult to define. If a person's body is no longer entirely biological, and their movements are augmented or controlled by external robotic systems, does that fundamentally alter their sense of self? What does it mean to be human if we begin to merge our brains with machines, enhancing our physical abilities and cognitive capabilities to superhuman levels?

These questions challenge traditional views of identity and selfhood. The human body, with its biological limitations, has long been seen as the defining feature of our identity. Yet, with Neuralink's potential to grant humans enhanced control over robotic limbs or exoskeletons, we enter a territory where the lines between the biological self and the technologically enhanced self begin to blur. If a person can perform actions that were previously impossible without the aid of machines, does that mean they are less human or more human? Is the essence of being human tied solely to our biology, or is it something more—something rooted in consciousness, intention, and the ability to interact with the world in meaningful ways?

The concept of "cybernetic superpowers" envisioned by Elon Musk could push this dilemma even further. As people gain access to enhanced physical abilities and cognitive functions through technology, they could become fundamentally different from those who choose not to augment themselves. This raises concerns about inequality—those who can afford such enhancements could gain significant advantages in both physical and intellectual domains. What happens to those who don't have access to these advancements? Will the divide between the augmented and non-augmented grow, leading to a new form of social stratification?

Furthermore, the issue of privacy becomes increasingly urgent in a world where our thoughts can directly control machines. If the brain is connected to technology, it's not just the body that becomes vulnerable to external influences, but the very essence of a person's mind. The potential for surveillance or manipulation of thoughts becomes a serious concern. Could a government, corporation, or malicious actor hijack a person's neural interface, controlling their actions or accessing their private thoughts? The ethical dilemma here is clear: how can we ensure that the use of BCIs like Neuralink is safe, secure, and transparent, without infringing on individual rights or autonomy?

Alongside these issues is the concern over consent. As these technologies become more integrated into society, especially in healthcare and rehabilitation, it's essential that individuals are fully informed about the risks and benefits. Neuralink's brain implants, for instance, require surgery and the potential for long-term side effects. Ensuring that users have a clear understanding of what the technology can—and cannot—do, as well as what risks they may be taking, is crucial for maintaining ethical standards in its development and implementation.

The most profound ethical question, however, revolves around the very essence of human existence. If our bodies are no longer purely biological, what happens to our understanding of what it means to be human? Are we still human if we enhance ourselves to the point where our minds are interlinked with machines, our bodies augmented with robotic limbs, or our cognitive abilities expanded beyond natural human limits? These questions touch on deep philosophical and existential issues, questioning the fundamental nature of humanity itself.

The integration of Neuralink and Tesla Optimus may one day make it possible for humans to transcend the limitations of our biology, but in doing so, we risk changing the very nature of who we are. As we approach a future where minds and machines are interconnected, we must carefully consider what it means to be human in a world where the body and mind are no longer strictly biological entities. The answers may not be clear-cut, but they are essential to guiding the ethical use of these transformative technologies.

In navigating these ethical challenges, it will be important for society to engage in open and transparent discussions about the potential risks and rewards of brain-computer interfaces and advanced robotics. Balancing the promise of enhanced human capabilities with the need to protect individual rights, maintain human dignity, and preserve the core values of humanity will be crucial as we move forward into this new technological era. Ultimately, it will require careful consideration, ongoing dialogue, and a commitment to ensuring that these innovations are used for the benefit of all, not just a select few.

As transformative as Neuralink and Tesla Optimus technologies may be, they also bring to light a significant concern: the potential for deepening inequality if these technologies remain out of reach for the majority of the population. The digital divide—disparities in access to technology based on factors such as income, geography, and education—has already shaped the way people interact with technology, and the introduction of advanced brain-computer interfaces and robotics could further exacerbate this divide.

If the cost of Neuralink implants and Tesla Optimus robots remains prohibitively expensive, only a small, wealthy segment of society would be able to access these life-changing technologies. This could create a new class of augmented individuals with enhanced physical, cognitive, and sensory capabilities, leaving the rest of the population behind. Over time, this gap could lead to a stark division between the "augmented" and "non-augmented," making it harder for those without access to the technology to compete in areas such as the job market, education, and even daily life. For example, in industries like healthcare, construction, and manufacturing, workers equipped with mind-controlled robots might be able to perform tasks far more efficiently and safely than their non-augmented counterparts. In this scenario, those who cannot afford the enhancements may find themselves at a severe disadvantage, further entrenching socioeconomic disparities.

The effects of this inequality would likely ripple through society, creating not only a technological divide but also a cultural one. As the rich and powerful augment themselves with advanced brain interfaces and robotics, they may enjoy a level of personal capability and productivity that is inaccessible to others. If such technologies are treated as a status symbol, it could lead to further social fragmentation, with an elite class of augmented humans separating themselves from the rest of society. In the worst-case scenario, the digital divide could result in a two-tiered society, where those who cannot afford such advancements face limited opportunities and diminished quality of life.

In addition to concerns about access and inequality, there are significant security risks associated with brain-controlled robotics that must be addressed before these technologies can be widely adopted.

Brain-computer interfaces like Neuralink and robotic systems like Tesla Optimus open up new frontiers in terms of human-machine interaction, but they also introduce vulnerabilities that could be exploited by hackers or malicious actors. If someone can control a robotic limb or a full-body exoskeleton with their thoughts, what happens if that system is hijacked by a third party? What if a hacker gains control of a person's neural interface and can manipulate their actions or, worse, cause harm?

The idea of brain-controlled robotics being hacked raises serious questions about the security of the brain itself. With so much data potentially flowing between the brain and external devices, the risks of exploitation become clear. A malicious actor could potentially access someone's neural data, including private thoughts or intentions, or take control of their robotic systems, leading to potential harm or violations of privacy. This is especially concerning in scenarios where mind-controlled robotics are used in sensitive environments, such as hospitals, workplaces, or even military applications.

To mitigate these risks, it will be crucial to develop robust security measures that protect users from unauthorized access or control. This may involve the use of advanced encryption, multi-factor authentication, and continuous monitoring of the brain-computer interface to detect any signs of tampering or intrusion. The challenge, however, is that the security systems in place must be as advanced as the technology itself—otherwise, vulnerabilities will be exploited. This also presents a unique dilemma: how can we ensure that the very technology designed to enhance human capabilities does not become a weapon in the hands of those who would misuse it?

Furthermore, the ability to remotely control robotic systems through thought raises concerns about the potential for malicious use of these technologies in areas like law enforcement or warfare. The notion of mind-controlled robots being deployed in conflict zones or surveillance situations brings with it a host of ethical and security issues. Could an individual's neural interface be hacked and used to carry out military strikes or other forms of control against their will? Could these technologies be used for mass surveillance, tracking people's movements or even intercepting their thoughts?

In response to these concerns, developers of Neuralink and Tesla Optimus will need to work closely with cybersecurity experts to create safeguards that are both effective and foolproof. It's essential that these technologies are designed with privacy, security, and ethical considerations at the forefront. This may involve not only technological solutions but also legal frameworks that protect individuals from misuse or exploitation of their neural data. Clear regulations will be necessary to ensure that the power of mind-controlled robotics is used responsibly and with respect for personal autonomy and privacy.

As exciting and transformative as these technologies are, their successful integration into society will depend on the ability to address these significant challenges. The digital divide must be narrowed, ensuring that these innovations are accessible to a broader population, and the risks associated with security and privacy must be carefully managed to prevent misuse. Only then can we begin to fully unlock the potential of Neuralink and Tesla Optimus, ensuring that their benefits are shared widely and safely across society.

As Neuralink and Tesla Optimus technologies evolve and become more prevalent, society will be faced with a range of complex challenges that must be addressed to ensure their responsible, ethical, and equitable integration into everyday life. These challenges span across ethical, social, economic, legal, and psychological dimensions, and they will require careful thought, dialogue, and collaboration between technologists, policymakers, ethicists, and the public.

One of the most immediate challenges is the ethical dilemma surrounding the enhancement of human capabilities through these technologies. Neuralink's brain-computer interface and Tesla Optimus's robotics open up the possibility of altering what it means to be human. While these advancements hold tremendous promise for improving the lives of those with disabilities, they also raise questions about where the line should be drawn between therapeutic interventions and enhancement. Should humans have the right to enhance their cognitive abilities, physical strength, or sensory perceptions using external technologies? What happens when individuals begin to rely on these technologies for personal or professional gain, potentially creating an unequal playing field?

The question of "what it means to be human" becomes especially pertinent as these technologies blur the boundaries between biology and machinery. With the potential for individuals to control robotic systems with their thoughts, or even augment their physical and cognitive abilities, society will need to grapple with profound questions about human identity. If a person can augment their physical capabilities with robotic limbs or enhance their mental capacities through brain-computer interfaces, at what point do they cease to be purely biological humans? What impact will this have on our concepts of agency, autonomy, and personal responsibility?

Beyond the ethical concerns, there are significant social challenges to consider. As these technologies become more accessible, the potential for societal division grows. As mentioned earlier, the "digital divide" could create an even greater gap between the rich and the poor, with those able to afford advanced implants or robotic enhancements gaining advantages in the workforce, education, and other areas. This divide could create a two-tier society where the augmented elite enjoy extraordinary capabilities while the rest of the population is left behind, without access to the life-changing benefits of these technologies.

Another social challenge lies in the potential for social and emotional isolation. If individuals rely heavily on robotic systems and mind-controlled interfaces for daily tasks, there's a possibility that they may become disconnected from human interaction, relying more on machines than on their own social networks. This shift could further erode traditional social bonds, leading to a society that is more technologically dependent but less connected emotionally. The balance between technological advancement and maintaining meaningful human relationships will be a critical area of concern as these technologies develop.

In terms of economics, the widespread adoption of neural and robotic technologies will bring about both opportunities and challenges. On the one hand, there will be new industries and job opportunities created by the development and application of these technologies. Industries such as healthcare, logistics, manufacturing, and even entertainment could benefit from a workforce augmented with mind-controlled robotics. The efficiencies and capabilities that come with these systems could lead to innovation in countless fields.

However, this also raises concerns about job displacement. As robots like Tesla Optimus take on more tasks, especially those requiring physical labor, there could be significant job losses in industries that traditionally rely on human workers. In particular, fields such as warehousing, assembly lines, and healthcare might see a reduction in the number of workers needed. The introduction of robotic assistants, controlled through Neuralink implants, could also challenge many professions that rely on physical and cognitive tasks, from doctors and surgeons to manual laborers. The displacement of workers could create significant social unrest, particularly if displaced workers are unable to access the training or resources needed to transition into new roles.

The economic implications will also depend on how the technology is distributed. If only a few companies or wealthy individuals are able to control and deploy Neuralink and Tesla Optimus technologies, it could result in monopolies or oligopolies, concentrating power and wealth in the hands of a select few. This could stifle competition and create an uneven playing field, where those with access to the technology continue to grow their influence while others struggle to keep up. Governments will need to step in to regulate these technologies, ensuring that their development does not lead to the consolidation of power or monopolistic practices.

In terms of privacy and security, these technologies introduce new vulnerabilities that must be addressed. Brain-computer interfaces and robotic systems are susceptible to hacking, which could have catastrophic consequences if personal thoughts, memories, or actions are compromised. The risks of cyber attacks on brain-controlled devices could lead to significant harm, both on an individual level and on a broader scale, especially if these systems are used in critical applications such as healthcare, military, or public infrastructure. Robust encryption and cybersecurity protocols will be essential to protect users' data, ensuring that their thoughts and actions remain private and secure.

Lastly, the psychological impact of these technologies on users will need to be considered. As people integrate neural interfaces and robotic systems into their lives, they may face new challenges related to self-perception, mental health, and dependence on machines. The experience of using mind-controlled devices to perform tasks or control robotic limbs could be overwhelming for some individuals, leading to psychological strain or a loss of personal autonomy. Additionally, users may experience challenges related to body image, as their perception of themselves shifts when interacting with machines or cybernetic enhancements. Therapies and support systems will be necessary to help individuals navigate these emotional and psychological hurdles.

In summary, as Neuralink and Tesla Optimus technologies become more prevalent, society will need to address a range of challenges, including ethical considerations, social inequality, economic disruption, privacy concerns, and psychological impacts. These challenges will require thoughtful policies, ethical frameworks, and global cooperation to ensure that the benefits of these technologies are distributed equitably and used responsibly. The potential of these advancements is undeniable, but their successful integration into society will depend on how well we navigate the complexities they introduce.

Chapter 7: The Road Ahead: Neuralink, Tesla, and the Future of Humanity

As Neuralink and Tesla Optimus continue to evolve, the future of both technologies holds exciting potential. With the ongoing trials and developments, the world is on the cusp of witnessing groundbreaking advancements that could dramatically reshape how we interact with machines and even how we experience the world itself.

Neuralink, having successfully developed the N1 brain-computer interface that allows individuals to control robotic limbs and digital devices with their thoughts, is already looking ahead to the next frontier in neural technology. One of the most ambitious projects currently under development is Neuralink's "Blindsight" initiative, aimed at restoring vision to the blind through advanced brain implants. This project could represent a significant leap forward in both the medical field and the realm of brain-machine interfaces.

Blindsight, as the name suggests, seeks to bypass the damaged parts of the eye or visual pathway in individuals who have lost their sight due to conditions such as retinal degeneration or optic nerve damage. By directly stimulating the visual cortex in the brain, Neuralink aims to provide these individuals with the ability to perceive visual stimuli, even in the absence of functioning eyes. Instead of relying on external devices like glasses or artificial eyes, which may only provide limited functionality, this brain-based approach has the potential to restore full or partial vision by transmitting visual data directly to the brain. In essence, the brain would "see" without needing the conventional physical mechanism of the eye.

While still in the early stages, Neuralink's progress in the realm of vision restoration is a testament to the company's ongoing drive to push the boundaries of what brain-computer interfaces can achieve. If successful, the Blindsight project could not only improve the lives of millions of people who are visually impaired but also redefine how we understand and treat sensory deficits. The implications of such a breakthrough extend far beyond healthcare—if the technology can be perfected, it could pave the way for new forms of sensory augmentation, where individuals can enhance their perception of the world beyond natural human capabilities.

As Neuralink works toward these ambitious goals, Tesla Optimus continues to make significant strides in the realm of humanoid robotics. Since its unveiling in 2021, Tesla's humanoid robot has captured the imagination of the public with its potential to revolutionize industries and improve the quality of life for people with disabilities or those requiring assistance with daily tasks. Tesla's engineers have been tirelessly refining the design and functionality of Optimus, with a focus on creating a robot that is both versatile and durable enough to be used in a wide range of real-world applications.

One of the most exciting developments for Tesla Optimus is the ongoing improvements to its dexterity and range of motion. With 22 degrees of freedom in its hands and further enhancements to its joints and body movements, Optimus is becoming increasingly capable of performing intricate tasks that require human-like precision. This is a critical factor for industries such as healthcare, where robots could be deployed to assist with delicate surgeries or rehabilitation processes, and manufacturing, where tasks requiring fine motor skills could be handled more efficiently by machines.

As these improvements continue, Tesla Optimus is poised to become an integral part of a wide range of industries. From industrial labor to healthcare, education, and even the arts, the potential applications of Optimus are limitless. The possibility of integrating Neuralink with Optimus could further elevate these capabilities, allowing individuals to control the robot with their thoughts and direct it to perform complex actions without the need for manual controls. This fusion of mind-controlled robotics could revolutionize the way people interact with machines, blurring the line between human and machine to a degree that we are only beginning to grasp.

Looking beyond the current developments, both Neuralink and Tesla Optimus are likely to continue exploring applications that go beyond simply restoring function to those with disabilities. As the technologies mature, we may see the creation of new possibilities for enhancing human capabilities. For instance, Neuralink's ability to enhance cognition or memory could open up new avenues for education, research, and even entertainment, while Tesla Optimus's increasing range of motion and adaptability could enable the creation of robots that assist in tasks traditionally performed by humans, from caregiving to creative endeavors like art and design.

One of the most exciting prospects for these technologies is the eventual convergence of artificial intelligence, brain-machine interfaces, and robotics into a cohesive system that can respond to human thought and intent with unparalleled precision. Imagine a world where humans are able to perform tasks that require significant physical strength, precision, or endurance, all through the use of thought-controlled robotics and augmented cognition. This could result in enhanced human performance, whether in athletics, surgery, or high-stakes manufacturing, where the boundaries of human potential are expanded beyond biological limitations.

The integration of AI into this ecosystem could further accelerate these possibilities. Neural networks, machine learning algorithms, and deep learning systems could be used to train the robots to understand and adapt to human commands more naturally, creating an even more seamless interaction between human thought and machine action. The synergy between human cognition and AI-powered robotics could lead to entirely new applications, enabling the creation of hyper-intelligent systems that work in tandem with human brains to solve complex problems.

Ultimately, the future of Neuralink and Tesla Optimus hinges on how effectively these technologies can be scaled and integrated into society. While there are still many hurdles to overcome—such as ensuring the safety, security, and accessibility of these systems—the potential for positive societal impact is enormous. As these technologies continue to evolve, we are likely to witness a revolution in the way humans interact with the world, pushing the boundaries of what we thought was possible in the realms of mobility, cognition, and human potential.

Tesla's ongoing work to refine Optimus, its humanoid robot, is focused on making it not just a tool for basic tasks, but a highly sophisticated system capable of performing complex, intricate tasks across a wide range of industries. One area where Optimus is being specifically developed for high-stakes work is healthcare. The goal is to create a robot that can assist doctors, surgeons, and healthcare workers in performing medical procedures with unprecedented precision. With its advanced dexterity—thanks to its 22 degrees of freedom in the hands and the ability to adapt its movements to the task at hand—Optimus is being trained to handle delicate, repetitive, or high-risk tasks that are currently done by human workers.

In surgery, for example, Optimus could assist surgeons by providing precise, steady movements that would reduce human error during intricate operations. The robot's hands can perform actions that require more control than the human hand can achieve, such as suturing, cutting, or manipulating surgical instruments in tight spaces. Moreover, its durability and flexibility could make it ideal for repetitive tasks in healthcare, such as sterilizing equipment, assisting with patient care, or even supporting rehabilitation exercises. With the ability to work around the clock without fatigue, Optimus could significantly enhance healthcare delivery, making life-saving procedures faster, more precise, and less dependent on human labor.

Another area where Optimus's capabilities are being honed is in space exploration. The potential for using humanoid robots like Optimus in the harsh environment of space could be transformative. Imagine astronauts aboard a spacecraft or at a space station being supported by robots that can handle routine maintenance, assist with complex experiments, and even perform tasks that are too dangerous for humans to do in zero gravity or extreme conditions. Optimus could be deployed to work alongside human astronauts in space exploration missions, enhancing their ability to perform their duties without the need to return to Earth for rest, recovery, or equipment repairs. In the long term, it's not hard to imagine a scenario where Tesla's robots are part of the crew that explores the Moon, Mars, or even farther into the cosmos, performing tasks like building infrastructure, mining resources, or conducting research in environments too hostile for humans.

The vision behind Neuralink and Tesla Optimus is not just about solving specific problems like mobility impairment or workforce efficiency. It is about fundamentally transforming what humans are capable of achieving, both individually and collectively. Neuralink, with its ability to interface directly with the brain, has the potential to enhance human cognition in ways that were once considered the stuff of science fiction. Beyond its applications in treating neurological disorders, Neuralink could one day allow people to dramatically increase their cognitive abilities—sharpening their focus, memory, and problem-solving skills, or even allowing them to "upload" and access information directly from the cloud. The possibilities for human enhancement through these technologies are virtually limitless. Neuralink could eventually become a tool for personal development on a scale that would enable people to learn and adapt faster, make more informed decisions, and extend their mental and physical capabilities beyond their natural limits.

At the same time, Tesla Optimus holds the potential to revolutionize industries by making human labor more efficient, reducing costs, and increasing safety in environments where humans are at risk. In areas like construction, manufacturing, and logistics, Optimus could take on heavy lifting, repetitive tasks, or work in dangerous environments—such as high-risk areas with hazardous materials, or in locations where human presence is limited or impractical. The integration of brain-computer interfaces with robots like Optimus could allow workers to control these machines with thought alone, increasing both productivity and precision while minimizing the physical strain that comes with manual labor.

But the potential impact of Neuralink and Tesla Optimus goes far beyond just business and industrial applications. These technologies could transform daily life in profound ways. For people with disabilities, the ability to control robotic limbs or even entire bodies through thought could drastically improve their quality of life, allowing them to perform tasks that were once unimaginable. In the future, enhanced mobility, cognitive abilities, and sensory experiences could be available to anyone who needs or wants them, potentially offering a new kind of equality—one based not on biology, but on access to technological enhancements.

The long-term vision for these technologies could create a world where humans and machines work in harmony, with advanced robotics performing much of the work that humans once did, and brain-machine interfaces enhancing human potential in ways we're only beginning to comprehend. In this world, human beings may be able to overcome the limitations of their physical bodies, and the role of the worker may evolve to one of more strategic oversight, creativity, and complex problem-solving.

The societal impact of such a shift would be far-reaching. While these technologies could dramatically improve quality of life for many, they could also disrupt entire industries and fundamentally change how we view work, personal capability, and human identity. Neuralink and Tesla Optimus could lead to a future where the distinction between what is human and what is robotic becomes increasingly difficult to define. As we augment our bodies and minds with advanced technologies, the very nature of what it means to be human will be challenged, forcing us to reconsider fundamental concepts like autonomy, agency, and what it means to live a fulfilling life.

In the long term, the integration of brain-computer interfaces and advanced robotics into society could lead to a world where the boundaries between human and machine blur. While this future holds immense promise, it will also require careful thought, planning, and consideration of the ethical, social, and economic implications of such profound changes. Whether this future becomes a utopia or a dystopia depends on how we navigate the challenges posed by these technologies today. The key will be ensuring that the benefits are distributed equitably, that the risks are managed responsibly, and that we remain conscious of the humanity at the heart of this technological revolution.

The collaboration between Neuralink and Tesla Optimus represents more than just the merging of two groundbreaking technologies—it marks the beginning of a new era in human evolution and technological advancement. The fusion of brain-computer interfaces with advanced humanoid robotics could redefine what it means to be human, unlocking possibilities that stretch the limits of our current understanding of biology, consciousness, and machine interaction.

At the heart of this collaboration is the potential to bridge the gap between the human mind and the mechanical world. Neuralink's N1 implant, which allows users to control robotic limbs and digital devices with their thoughts, already shows the transformative potential of brain-computer interfaces. But when combined with Tesla Optimus—robots designed to replicate human biomechanics and carry out complex tasks with precision and dexterity—the technology moves beyond simple mobility aids or medical devices. It begins to represent a future where humans and machines work together in ways that were once confined to the realm of science fiction.

This collaboration could drastically alter the course of healthcare, giving people with physical disabilities the ability to regain lost capabilities and, in some cases, surpass what was once possible for a human body. The possibilities for rehabilitation, assistive technology, and prosthetics are limitless. But beyond the medical field, the integration of mind-controlled robotics could radically change how industries operate, making them more efficient, cost-effective, and safe. Tasks that are physically demanding or too dangerous for humans—such as lifting heavy objects, working in hazardous environments, or performing high-risk surgeries—could be taken over by robotic systems, operated directly by the thoughts of a human controller.

The long-term implications of this collaboration are even more profound. Neuralink and Tesla Optimus could lead to an era of human enhancement, where physical and cognitive limitations are no longer barriers to achieving our goals. Enhanced strength, precision, cognitive abilities, and even sensory perception could become the new normal. Instead of relying on external aids like glasses, hearing aids, or wheelchairs, individuals could use brain-machine interfaces to control advanced robotic systems that function seamlessly as extensions of their own bodies. The concept of a "cybernetic superhuman" may soon move from theoretical to practical, allowing people to perform extraordinary feats that were previously unimaginable.

On a societal level, this collaboration could profoundly reshape the structure of work and the economy. As more tasks are automated, there will likely be a shift in how labor is perceived and valued. Jobs that rely on physical strength, repetitive tasks, or even intricate craftsmanship could be performed by robots, freeing humans to focus on higher-level creative, strategic, or intellectual pursuits. This could lead to a more equitable distribution of labor, where human potential is leveraged to its fullest extent, while robots handle the heavy lifting. The integration of mind-controlled robotics into daily life could also create new opportunities for leisure, education, and personal development, allowing individuals to pursue passions and interests that were once out of reach.

But these advancements also raise critical questions. What happens when the line between human and machine becomes increasingly blurred? As people enhance their abilities with these technologies, will society still define them as human, or will they be seen as something else—part biological, part artificial? The ethical implications of human enhancement will need to be navigated carefully, ensuring that these technologies are used responsibly and equitably. The very definition of what it means to be human could be rewritten, challenging long-held assumptions about identity, autonomy, and the role of technology in our lives.

The collaboration between Neuralink and Tesla Optimus is not just about improving individual lives—it has the potential to change the very fabric of humanity. If these technologies can be integrated successfully into society, they could help solve some of the most pressing challenges facing the world today, from healthcare access to labor shortages, from aging populations to space exploration. This fusion of mind and machine might be the key to unlocking new levels of human potential, making a future where humans and machines coexist harmoniously not just a dream, but a reality.

In the grand sweep of history, the convergence of Neuralink's brain-computer interfaces and Tesla Optimus's robotics might one day be seen as the turning point where humanity began to transcend its physical limitations and step into a new era—one where technology and the human mind work together to create possibilities that were once beyond our imagination. This is not just a technological evolution; it is the dawn of a new chapter in the story of human progress. The future, as envisioned by Elon Musk and the teams at Neuralink and Tesla, could be one where humanity evolves alongside its creations, not just adapting to technology, but advancing together with it.

Conclusion

As we look back on the journey through this transformative narrative, we are left with a profound understanding of how the fusion of Neuralink's cutting-edge brain-computer interface technology and Tesla's advanced humanoid robots like Optimus is reshaping the world as we know it. These innovations are not just milestones in the world of robotics and neural science; they are gateways to a future where the boundaries between human abilities and machines blur in ways that were once unimaginable.

The promise of Neuralink's N1 implant, which allows thoughts to control digital devices and robotic limbs, opens up new possibilities for people with disabilities, offering them unprecedented levels of independence and mobility. As we've seen, the potential to expand this technology into other realms, from enhanced cognitive abilities to sensory restoration, could dramatically improve quality of life for millions across the globe. Meanwhile, Tesla

Optimus is pushing the limits of humanoid robotics, creating machines that not only replicate human biomechanics but also possess the flexibility, precision, and dexterity to perform complex tasks across industries like healthcare, space exploration, and manufacturing. The combination of these two technologies could create a future where both mind-controlled robotics and human enhancement are integrated into everyday life, revolutionizing how we interact with the world and each other.

This partnership between Neuralink and Tesla is more than a technological leap forward; it represents a profound shift in how we understand human potential. Whether it's restoring the ability to move or enabling the mind to perform tasks that once required physical effort, these innovations are poised to transform healthcare, industries, and even the way we perceive human limitations. With robots like Optimus taking over mundane, dangerous, or repetitive tasks, humans will be free to focus on more creative, strategic, and meaningful work—leading to a world where efficiency and human potential reach new heights.

Yet, with all the promise comes responsibility. As these technologies advance, we must ensure they are used ethically, equitably, and for the collective good. The questions of identity, accessibility, and security are critical, and it will be up to us to navigate these challenges as we move into this new era. We are standing on the precipice of something monumental—an opportunity to redefine what it means to be human in the age of advanced machines and mind-controlled robotics.

The future is unfolding before us at an astonishing pace. Neuralink and Tesla Optimus are just the beginning of a journey toward unprecedented human enhancement, robotic integration, and mind-machine collaboration. To fully embrace what lies ahead, we must stay informed, stay curious, and remain open to the possibilities that these technologies offer. The changes we're witnessing today will not only impact the lives of individuals with disabilities or those working in cutting-edge industries but will ripple through every aspect of society, touching healthcare, education, work, and even our fundamental understanding of what it means to be human.

As we look to the future, the most important thing we can do is prepare ourselves—mentally, emotionally, and socially—for the revolutionary changes that are already on the horizon. Embrace the power of innovation, and let us be part of the world that adapts and thrives in this new age of technology. Stay curious, stay informed, and be ready for the incredible journey that awaits. The future is here, and it's just getting started.

www.ingramcontent.com/pod-product-compliance
Lightning Source LLC
Chambersburg PA
CBHW071037240526
45469CB00006BD/2238